Abdelhafid Mimouni

Ferredoxin and Ferritin: An Evolving Perspective

Abdelhafid Mimouni

Ferredoxin and Ferritin: An Evolving Perspective

ScienciaScripts

Imprint

Any brand names and product names mentioned in this book are subject to trademark, brand or patent protection and are trademarks or registered trademarks of their respective holders. The use of brand names, product names, common names, trade names, product descriptions etc. even without a particular marking in this work is in no way to be construed to mean that such names may be regarded as unrestricted in respect of trademark and brand protection legislation and could thus be used by anyone.

Cover image: www.ingimage.com

This book is a translation from the original published under ISBN 978-620-6-72229-8.

Publisher:
Sciencia Scripts
is a trademark of
Dodo Books Indian Ocean Ltd. and OmniScriptum S.R.L publishing group

120 High Road, East Finchley, London, N2 9ED, United Kingdom
Str. Armeneasca 28/1, office 1, Chisinau MD-2012, Republic of Moldova, Europe
Printed at: see last page
ISBN: 978-620-8-12725-1

Ferredoxin and Ferritin: An Evolving Perspective

Author : Dr. Abdelhafid Mimouni

An independent researcher in bioinorganic chemistry, Dr Mimouni is an expert in macromolecular synthesis and characterisation. He obtained his PhD in Chemistry from the University of Paris XII in 1997, after a Diplôme des Études Approfondies in Bioinorganic Systems from the University of Paris XI in 1993, where he also obtained his Licence and Maîtrise in Chemistry.

Summary

Iron regulation is vital for living organisms, with distinct mechanisms in bacteria and humans. Bacteria use ferredoxin for metabolic processes, while ferritin and transferrin manage iron storage and transport in humans. Pathologies linked to ferritin abnormalities underline the importance of balanced iron regulation. A better understanding of these mechanisms could improve treatments and inspire new therapies.

Book outline :

Chapter 1 Introduction to Ferredoxin and Ferritin :

Iron management is crucial to the functioning of biological systems, and the proteins involved in this process vary considerably between bacteria and multicellular organisms such as humans . Two emblematic proteins in this context are ferredoxin and ferritin. Although these molecules play essential roles in iron metabolism, they do so in distinct ways depending on their specific structures and functions.

Ferredoxin

Ferredoxin is a small, iron-sulphur-rich protein found mainly in bacteria, algae and plants. Its structure is characterised by the presence of iron-sulphur clusters, which are electron transfer centres in redox reactions. These clusters enable ferredoxin to participate in essential biological processes such as photosynthesis and nitrogen fixation. In bacteria, ferredoxin plays a key role in biochemical reactions involving electron transfer, contributing to energy conversion and the reduction of various substrates.

The main function of ferredoxin is therefore to act as an electron transporter in metabolic pathways, facilitating enzymatic reactions that are crucial for the survival and growth of micro-organisms. Unlike iron storage proteins, ferredoxin is not directly involved in storing or regulating iron, but rather in mediating its redox transfer in metabolic processes.

Ferritin

Ferritin, on the other hand, is an iron storage protein that is ubiquitous in eukaryotes, including humans. Its structure consists of a spherical protein shell that encapsulates iron atoms in the form of ferric minerals. This complex enables ferritin to store large quantities of iron in a non-toxic and bioavailable manner, thus playing a crucial role in maintaining iron balance in the body.

Ferritin is particularly important in the regulation of iron, as it releases and captures iron according to physiological needs. It ensures that iron is available for biological processes such as haemoglobin synthesis and enzymatic reactions. Proper management of iron by ferritin is vital in preventing pathological conditions such as iron-deficiency anaemia or haemochromatosis, a genetic disease characterised by excessive accumulation of iron in tissues.

Differences in Iron Storage and Transport Mechanisms

The differences between ferredoxin and ferritin reflect their distinct roles in iron management. While ferredoxin is involved in the transfer of electrons during redox reactions, ferritin specialises in the secure and regulated storage of iron. Iron storage and transport mechanisms thus vary according to the biological needs of organisms and their specific molecular structures.

Ferredoxin has no direct role in iron storage, but is essential for reactions that use iron as a cofactor. Ferritin, on the other hand, is an adaptive protein that adjusts iron levels according to metabolic needs, providing a valuable reserve for vital biological processes.

By comparing these two proteins, we can better understand how different forms of life have evolved distinct strategies for managing one of the most essential elements for life: iron.

Chapter 2 Analysis of Ferritin-Associated Diseases

Ferritin is a key protein in the management of iron in the human body. As the main reservoir of iron, it plays a crucial role in the storage and release of iron according to physiological needs. Abnormalities in the function or regulation of ferritin can lead to a variety of pathologies, which can have significant consequences for human health. This chapter explores the genetic diseases associated with ferritin deficiency, such as haemochromatosis and iron absorption disorders, as well as the clinical and pathophysiological impacts of these deficiencies.

Genetic Diseases Associated with Ferritin Deficiency

1. **Haemochromatosis**

   Haemochromatosis is a genetic disease that leads to excessive iron overload in the body. The condition results mainly from mutations in the HFE gene, which is responsible for regulating iron absorption in the intestine. The most common mutations, such as C282Y and H63D, disrupt the regulatory mechanism, leading to increased iron absorption.

   This iron overload is deposited in various organs, including the liver, pancreas and heart, and can lead to serious complications such as cirrhosis of the liver, diabetes mellitus (due to damage to the pancreas) and cardiomyopathy. Clinical symptoms include fatigue, joint pain and hyperpigmentation of the skin. Early diagnosis is crucial and is based on genetic and biochemical tests. Management of haemochromatosis generally involves therapeutic bloodletting to reduce iron levels and prevent organ damage.

2. **Iron absorption disorders**

Iron absorption disorders can also result from ferritin deficiency. Ferritin deficiency is often associated with conditions such as iron deficiency anaemia, where a lack of iron results in insufficient production of red blood cells. This condition can be caused by insufficient ferritin production, excessive iron loss, or inefficient iron absorption.

Symptoms of ferritin deficiency include fatigue, pallor and cognitive impairment. Therapeutic approaches focus on iron supplementation and management of the underlying causes. Genetic mutations affecting the synthesis or regulation of ferritin can also lead to specific forms of iron metabolism disorders.

Clinical and Pathophysiological Impacts of Ferritin Deficiencies

Ferritin deficiencies have a significant impact on health, both clinically and pathophysiologically.

- **Clinical impact:** Pathologies associated with iron overload, such as haemochromatosis, can cause serious organ damage and affect quality of life. Patients can suffer from chronic fatigue, joint pain and major metabolic complications. On the other hand, ferritin deficiency can lead to symptoms such as anaemia, with notable side-effects on daily performance and cognitive ability.

- **Pathophysiological impact:** Excess iron causes significant oxidative stress, leading to cellular damage and secondary diseases, including cancer and cardiovascular disease. In the case of deficiency, the absence of ferritin results in a failure in the storage and transport of iron, leading to anaemia and disturbances in the synthesis of red blood cells.

Analysis of these pathologies highlights the importance of precise regulation of iron in the body. Monitoring ferritin levels and appropriate management are essential to optimise health and prevent the serious complications associated with abnormal iron metabolism.

Chapter 3 Comparison of Iron Regulation Mechanisms

Iron regulation is essential for maintaining the balance of this vital element in living organisms. Bacteria and humans adopt distinct strategies for regulating and storing iron, due to their different biological needs and structures. This chapter compares the mechanisms of iron regulation and storage in these two groups, and examines why some iron-related pathologies are specific to humans and not to bacteria. We will also explore specific examples of bacteria using ferredoxin in their iron metabolism.

Regulation and Storage of Iron in Bacteria

As single-celled organisms, bacteria need to regulate iron efficiently in order to survive in environments that are often iron-poor. Here are some of the mechanisms used by bacteria:

1. **Iron Capture Systems:** Bacteria often secrete siderophores, molecules capable of binding iron with high affinity, facilitating its absorption. For example, *Escherichia coli* produces the siderophore enterobactin, which binds iron and transports it to the bacterial cell.

Iron Capture Systems

Because bacteria are often exposed to environments where iron is present in very limited quantities, they have developed sophisticated mechanisms for capturing and transporting this vital element. One of the most efficient systems used by bacteria is the production of siderophores, which are specialised molecules capable of binding iron with very high affinity. Here is a detailed explanation of this process and a specific example using *Escherichia coli.*

How Siderophores work

Siderophores are small, highly selective molecules that bind to iron (Fe^{3+}) with extremely high affinity. This high affinity is crucial, as iron in the environment is often in the form of ferric iron (Fe^{3+}), which is poorly soluble and difficult for cells to capture. Here's how siderophores work:

Secretion: Bacteria synthesise siderophores in their cytoplasm and secrete them into the environment. These siderophores bind to the iron present in the surrounding environment.

Iron capture: Siderophores have functional groups (such as hydroxyls and carboxyls) that bind specifically to ferric iron, forming a siderophore-iron complex. This complex is very stable, allowing efficient capture of iron even at low concentrations.

Transport to the cell: Once the siderophore has captured the iron, it binds to specific receptors on the surface of the bacterial cell. The siderophore-iron complex is then internalised by active transport or receptor-mediated binding mechanisms, allowing the iron to enter the cell.

Iron release: Inside the cell, the siderophore-iron complex is often dissociated by chemical reductions or pH changes, releasing the iron for use in various metabolic processes.

Example of *Escherichia coli* and Enterobactin

A typical example of an iron uptake system is seen in *Escherichia coli*, a Gram-negative bacterium that is commonly studied. *E. coli* produces a siderophore called **enterobactin**, which plays a crucial role in iron uptake.

1. **Synthesis and secretion:** *E. coli* synthesises enterobactin in the cytoplasm and excretes it into the environment. Enterobactin consists of three N-hydroxyethylene triacetic diamine (NHE3) units linked together, each capable of binding ferric iron.

2. **Iron capture:** Enterobactin binds with very high affinity to ferric iron available in the extracellular environment, forming a stable complex. This high affinity enables *E. coli* to capture iron even when it is present in low concentrations.

3. **Transport and internalisation:** Once the siderophore-iron complex has been formed, it binds to a specific receptor on the surface of *E. coli*, called FepA (ferroportin A). The complex is then internalised into the cell by a process called endocytosis.

4. **Use of iron:** Inside the cell, iron is released from enterobactin by specific chemical reductions or pH changes. Iron is then used for

various cellular functions, such as DNA synthesis, cellular respiration and enzymatic reactions.

Importance of siderophores

Siderophores are essential for the survival of bacteria in iron-poor environments. They allow bacteria to :

- **Acquiring iron:** Siderophores increase the availability of iron, an essential nutrient, even in conditions of deficiency.

- **Avoiding competition:** By producing their own siderophores, bacteria can obtain iron before other micro-organisms in the same environment.

- **Surviving in hostile environments:** Siderophores enable bacteria to colonise specific ecological niches, including internal or hostile environments such as animal tissue or industrial environments.

In summary, siderophores are key molecules that enable bacteria to capture iron efficiently in often unfavourable environments. Their production and function, as observed with enterobactin in *Escherichia coli*, illustrate the ingenuity of the mechanisms bacteria use to meet their iron requirements.

2. **Iron storage:** Bacteria use iron storage proteins such as ferritin and bacterioferritin. A notable example is *Bacillus subtilis*, which stores iron using bacterioferritin to regulate intracellular iron levels.

3. **Regulation:** Bacteria adjust the expression of genes involved in iron absorption and storage in response to available iron levels. *Pseudomonas aeruginosa*, for example, regulates the production of siderophores according to the availability of iron in its environment.

4. **Ferredoxin:** Ferredoxin is an iron-containing protein involved in electron transfer in various metabolic processes. In bacteria such as *Clostridium acetobutylicum*, ferredoxin plays a crucial role in nitrogen fixation and fermentation, using iron to catalyse these reactions.

Ferredoxin: Function and Metabolic Role

Ferredoxin is an iron-containing protein that is essential for electron transfer during various metabolic processes. This protein plays a key role in several biochemical pathways, particularly in bacteria, where it is involved in crucial reactions such as nitrogen fixation and fermentation.

Structure and properties of Ferredoxin

Ferredoxin is characterised by its structure containing iron-sulphur clusters, which are complexes of iron and sulphur arranged in a specific pattern. These iron-sulphur clusters can vary in size and composition, but their main function is to facilitate the transfer of electrons between molecules. Ferredoxin's ability to accept and donate electrons is crucial to its enzymatic roles.

Role in nitrogen fixation

In diazotrophic bacteria, such as *Clostridium acetobutylicum*, ferredoxin plays a fundamental role in nitrogen fixation. Nitrogen fixation is the process by which nitrogen gas (N_2) in the atmosphere is converted into ammonia (NH_3), a form that can be used by plants and micro-organisms. This process is catalysed by the enzyme nitrogenase, which uses ferredoxin molecules to transfer the electrons needed to reduce the nitrogen.

In this reaction, ferredoxin supplies electrons to nitrogenase, enabling atmospheric nitrogen to be converted into ammonia. Nitrogenase is a complex enzyme that requires a significant amount of energy and electron transfer to accomplish this conversion. Ferredoxin, with its iron-sulphur clusters, plays a key role in facilitating this electron transfer because of its ability to switch easily between oxidation states.

Role in Fermentation

Fermentation is a metabolic process by which bacteria convert organic substrates into final products, often to generate energy in the absence of oxygen. In *Clostridium acetobutylicum*, ferredoxin is involved in the fermentation of various substrates, including sugars, to produce organic acids, alcohols and other compounds.

Ferredoxin acts here by transferring electrons to enzymes involved in fermentation pathways. For example, in butyric fermentation, ferredoxin transfers electrons to butyrate kinase and other enzymes, facilitating the conversion of substrates into butyric acid and other fermentation products.

Importance of Ferredoxin in Metabolic Processes

Ferredoxin's ability to transfer electrons is essential for many metabolic processes:

- **Energy efficiency:** Ferredoxin enables rapid, efficient electron transfer, facilitating biochemical reactions that are crucial to the survival and growth of bacteria.
- **Compound reduction:** By providing electrons for reduction reactions, ferredoxin is involved in the transformation of various compounds, making these biochemical processes possible.
- **Metabolic adaptation:** The presence and activity of ferredoxin enable bacteria to adapt to different environments and metabolise different substrates, contributing to their functional and ecological diversity.

In summary, ferredoxin is a vital protein that plays a central role in electron transfer in various metabolic processes, including nitrogen fixation and fermentation, using iron to catalyse these reactions. Its function is essential for the metabolism of bacteria such as *Clostridium acetobutylicum*, underlining the importance of this protein in biological cycles and biochemical processes.

Iron Regulation and Storage in Humans

In humans, iron regulation is more complex because of the need to maintain appropriate levels of iron in a multicellular organism:

1. **Iron absorption:** Iron is mainly absorbed in the small intestine. The iron transporter, ferroportin, plays a key role in the passage of iron from the inside of intestinal cells into the blood. Iron absorption is regulated by the production of hepcidin, a liver hormone that controls the release of stored iron into macrophages and the liver.

2. **Iron storage:** In humans, ferritin is the main iron storage mechanism. It is found mainly in the liver, spleen and bone marrow. In addition to ferritin, haemosiderin is another form of iron storage, generally observed in cases of iron overload.

3. **Regulation:** Iron regulation in humans is influenced by hormonal and genetic mechanisms. Hepcidin plays a central role in regulating iron absorption and storage by modulating ferroportin activity. Factors such as inflammation and body iron requirements can influence hepcidin levels and, consequently, iron availability.

Specificity of iron-related diseases

Certain pathologies associated with iron disorders are specific to humans and not observed in bacteria due to fundamental differences in biological systems:

1. **Haemochromatosis:** This genetic disease, characterised by excessive iron overload, is specifically human. Mutations in genes regulating iron absorption, such as HFE, are unique to humans and are not observed in bacteria. The mechanisms by which iron is regulated in bacteria do not lead to similar pathologies because their regulation is adaptive and often simpler.

2. **Iron deficiency anaemia:** Although bacteria can be iron deficient, iron deficiency anaemia is a clinical disorder specific to humans. This condition is linked to complex factors, including the regulation of ferritin and the demand for iron for red blood cell production, which are not directly comparable to the mechanisms of iron regulation in bacteria.

3. **Iron overload:** Humans are subject to iron overload conditions such as haemochromatosis and haemosiderosis, specific pathologies due to the complexity of the iron regulatory system and the absence of equivalent mechanisms in bacteria.

In conclusion, comparison of iron regulation and storage mechanisms between bacteria and humans reveals significant differences due to the complexity and distinct biological needs of these organisms. The examples of bacteria using ferredoxin illustrate the diversity of metabolic strategies for managing iron. The pathologies associated with iron disorders reflect these differences and show how unique evolutionary mechanisms can influence the diseases specific to each group.

Chapter 4 Evolution and Adaptation: Iron Storage and Regulation Mechanisms

Evolution has shaped various biological mechanisms, including those involved in the storage and regulation of iron. Iron is an essential element for all living organisms, playing a crucial role in biological processes such as cellular respiration, photosynthesis and DNA synthesis. However, the way in which iron is stored and regulated varies considerably between different groups of organisms, reflecting evolutionary adaptation to specific needs and varied environments.

Evolution of Iron Storage and Regulation Mechanisms

1. Adaptation in bacteria

Iron storage and regulation mechanisms in bacteria have evolved to meet the specific challenges of their environment. Bacteria living in iron-poor environments have developed unique strategies for capturing and using this essential nutrient:

- **Siderophores:** As mentioned above, bacteria produce siderophores to capture iron from environmental sources. This adaptation allows them to thrive in iron-limited environments, using high-affinity molecules to extract ferric iron (Fe^{3+}) from their environment.
- **Intracellular storage:** Some bacteria use iron storage proteins, such as bacterianin, to store iron inside the cell. These proteins enable bacteria to store iron during periods when it is abundant and release it when levels are low.

- **Regulatory systems:** Bacteria have developed sophisticated regulatory systems to control the expression of genes involved in iron uptake and utilisation. For example, the Fur (ferric uptake regulator) protein regulatory system regulates the transcription of genes linked to iron uptake in response to available iron levels.

2. Adaptation in eukaryotes :

In eukaryotes, iron storage and regulation mechanisms are also adapted to their specific needs and environments:

- **Ferritin and haemosiderin:** In animals, iron is mainly stored in the form of ferritin and haemosiderin in the cells. Ferritin is a protein that stores iron in a soluble, bio-available form, while haemosiderin is an insoluble form of iron storage, often associated with pathologies such as haemochromatosis.
- **Transport systems:** Eukaryotes use iron transport proteins such as transferrin, which binds iron and transports it in the blood to the cells that need it. Regulation of transferrin is essential for maintaining iron balance and preventing iron overload or deficiency.
- **Absorption mechanisms:** Eukaryotes, like humans, have developed iron absorption mechanisms in the intestine, where dietary iron is absorbed and regulated according to body

requirements. Proteins such as ferritin and hepcidin regulate this absorption to avoid excess iron.

Chapter 5 Evolution of Genetic Pathologies and their Human Specificity

1. Pathologies associated with ferritin deficiency :

Pathologies linked to ferritin deficiency in humans, such as haemochromatosis and iron absorption disorders, are often the result of genetic mutations that affect iron regulation. These diseases are not observed in bacteria, and there are several reasons for this specificity:

- **Metabolic complexity:** Iron-related human pathologies are often due to defects in specific iron storage and regulatory proteins, such as ferritin, transferrin and hepcidin. These proteins play complex roles in iron regulation, and mutations can cause significant imbalances, leading to disease.

- **Pathway specificity:** Iron storage and regulation mechanisms in bacteria and eukaryotes are adapted to their different physiological needs. Bacteria have less complex iron capture and storage systems than eukaryotes. Genetic mutations affecting iron regulation in eukaryotes may have no direct analogue in bacteria.

- **Evolutionary differences:** Evolution has led to iron regulation mechanisms that are specific to eukaryotes, including humans. These mechanisms are often more sophisticated due to the complexity of biological functions and more varied metabolic needs. Iron-related pathologies in humans often result from fine deregulation of these complex systems, whereas bacteria have not evolved to develop such pathologies.

2. Examples of Pathologies Specific to Humans :

- **Haemochromatosis:** A genetic disease characterised by excessive iron absorption, often due to mutations in the HFE gene. This overload of iron can lead to damage to organs such as the liver, heart and pancreas.

- **Iron deficiency anaemia:** Resulting from iron deficiency, this condition can be caused by absorption disorders, chronic bleeding, or mutations affecting iron regulation. Symptoms include extreme fatigue and reduced production of red blood cells.

In summary, iron storage and regulation mechanisms have evolved to meet the specific needs of different groups of organisms. Bacteria and eukaryotes, such as humans, have developed unique adaptations to manage iron in their respective environments. Genetic pathologies linked to iron in humans, such as haemochromatosis and iron deficiency anaemia, illustrate the complexity of iron regulation systems and the specificity of these diseases for eukaryotes. These diseases are not seen in bacteria, partly because of evolutionary differences in iron regulation mechanisms.

Chapter 6 Applications and Research Perspectives

Studying the mechanisms of iron storage and regulation in bacteria and eukaryotes, and the associated pathologies, has important implications for biomedical research and potential treatments. A thorough understanding of these mechanisms may lead to new strategies for the treatment of iron-related diseases and open up avenues for future comparative research.

Implications for Biomedical Research

1. Advances in the Treatment of Iron-Related Diseases :

The knowledge acquired about iron regulation can directly influence the development of treatments for diseases such as haemochromatosis and iron deficiency anaemia. For example:

- **Targeted therapies:** A better understanding of the mechanisms by which iron is regulated could lead to the design of targeted therapies that specifically modulate the proteins involved in iron absorption and storage. Agents that regulate the production of hepcidin or other proteins involved in iron metabolism could offer therapeutic options for treating iron-related disorders.

- **Development of new drugs:** Research into siderophores and iron storage proteins in bacteria could inspire the development of new drugs. For example, compounds that mimic the action of siderophores could be used to interact with iron transport proteins in the human body and treat iron deficiency or excess.

- **Biomarkers:** Understanding differences in iron storage mechanisms could lead to the discovery of biomarkers for more accurate diagnosis of iron-related diseases. Specific biomarkers could facilitate early detection and more effective management of conditions such as haemochromatosis and iron absorption disorders.

2. Applications in Medicine :

Advances in our understanding of iron regulation mechanisms have potential applications in medicine:

- **Gene therapies:** Research into genetic mutations affecting iron regulation could lead to gene therapy approaches to correct the defects responsible for iron-related pathologies.
- **Chelating drugs:** Knowledge of the biochemistry of siderophores may inspire new iron chelators that could be used to treat iron overload, particularly in conditions such as haemochromatosis.

Avenues for Future Comparative Research

1. Comparison of Regulatory Mechanisms :

In-depth comparative studies of iron regulation mechanisms in bacteria and eukaryotes can reveal fundamental similarities and differences:

- **Molecular mechanisms:** Comparing iron storage and regulation proteins, such as ferritin, transferrin and siderophores, can provide information on evolutionary adaptations and strategies specific to each group of organisms.

- **Evolutionary adaptations:** An analysis of evolutionary adaptations in iron regulation mechanisms could help us to understand how different organisms have developed strategies to cope with iron-related challenges, particularly in varied environments.

2. Functional studies :

Functional research can explore how iron regulatory mechanisms affect cell biology and the pathophysiology of disease:

- **Animal models:** The use of animal models to study genetic mutations affecting iron regulation can provide insights into the clinical and pathophysiological impact of these mutations.
- **Structural biology approaches:** Structural analysis of proteins involved in iron metabolism, such as storage and transport proteins, can reveal mechanisms of action and potential sites for therapeutic intervention.

3. Technological innovations :

Technological innovations could improve iron research:

- **Advanced spectroscopy:** The use of advanced spectroscopy techniques, such as X-ray absorption spectroscopy (XAS), could provide detailed information on iron complexes in different biological contexts.
- **Bioinformatics:** Bioinformatics tools can help analyse the sequences and structures of proteins involved in iron regulation to identify new therapeutic targets.

In summary, our knowledge of iron regulation and associated pathologies opens up numerous opportunities for biomedical research and the development of new treatments. Comparative research into the mechanisms of iron regulation in bacteria and eukaryotes, as well as functional and technological studies, offer promising avenues for future investigations.

Biomedical research benefits greatly from the comparative study of these mechanisms, offering prospects for the development of more targeted treatments and management strategies for iron-related diseases. Exploring the similarities and differences in iron regulation between bacteria and humans opens up promising avenues for future investigations. By integrating innovative approaches and advanced technological tools, researchers can advance our understanding of iron-related pathologies and improve therapeutic interventions. Interdisciplinary collaboration and the application of fundamental discoveries to clinical practice will continue to play a crucial role in managing iron disorders and promoting overall health.

Chapter 7: Differences in Iron Regulation and Storage: Comparison between Bacteria and Humans

Introduction

Iron is an essential element for all living organisms, playing a crucial role in various biological processes, including oxygen transport, cellular respiration and DNA synthesis. However, the way in which iron is regulated and stored varies considerably between bacteria and humans, reflecting specific evolutionary adaptations to their environments and needs. This chapter explores these differences and analyses why certain iron-related genetic pathologies specifically affect humans and not bacteria.

Iron regulation and storage mechanisms

1. Iron regulation in bacteria

Because of their diversity and varied environments, bacteria have developed distinct mechanisms for capturing and using iron. Two notable examples are *Escherichia coli* and *Clostridium acetobutylicum*.

- **Siderophores**: Many bacteria, including *E. coli*, secrete siderophores, molecules that specialise in capturing free iron from the environment. Enterobactin, a siderophore produced by *E. coli*, has a very high affinity for iron, enabling the bacteria to absorb this mineral even in low concentrations. Once the iron has been captured, it is transported to the bacterial cell, where it is used for various metabolic functions.

- **Ferredoxin**: *Clostridium acetobutylicum* uses ferredoxin, an iron-containing protein, to catalyse key reactions such as nitrogen fixation and fermentation. Ferredoxin plays a crucial role in these processes by transferring electrons, facilitating the reduction of compounds and the production of energy.

2. Iron regulation in humans

In humans, iron regulation is more complex and involves several key proteins, including ferritin and transferrin.

- **Ferritin**: Ferritin is an intracellular protein responsible for storing iron. It stores iron in a non-toxic way and releases it according to cellular needs. A deficiency in ferritin can lead to iron-deficiency anaemia, while an excess can cause pathologies such as haemochromatosis, an excessive accumulation of iron in the organs.
- **Transferrin**: Transferrin is a plasma protein that transports iron in the blood. It binds to iron and transports it to the tissues where it is needed, thus regulating iron absorption and use.

Comparison of mechanisms

The mechanisms for regulating and storing iron in bacteria and humans differ considerably:

- **Bacterial flexibility**: Bacteria such as *E. coli* and *C. acetobutylicum* show great flexibility in their ability to capture and use iron, enabling them to survive in varied and often iron-deficient environments. Siderophores and ferredoxin enable them to function effectively even when iron availability is limited.

- **Complexity in mammals**: In contrast, humans depend on a more complex system for regulating iron. Ferritin and transferrin ensure precise control of iron, necessary to maintain the balance between absorption, storage and use. Genetic disturbances affecting these proteins can lead to serious illnesses because of the need for rigid control of iron in the human body.

Iron-related genetic disorders

Some iron-related genetic diseases, such as haemochromatosis, only occur in humans because of their specific iron regulation mechanisms:

- **Haemochromatosis**: This genetic disease leads to an excessive accumulation of iron in the organs, causing tissue damage and organ failure. Genetic mutations affecting proteins involved in iron regulation, such as ferritin, can lead to this condition.

- **Iron deficiency anaemia**: Ferritin deficiency can also lead to iron deficiency anaemia, a condition characterised by iron deficiency and reduced red blood cell production.

- **Genetic Evolution and Adaptation of Iron Regulation Mechanisms**

- The evolution of iron regulation and storage mechanisms in humans and bacteria has been shaped by environmental pressures and specific physiological needs. Gene mutations associated with disorders of iron metabolism, such as those causing haemochromatosis, illustrate how changes in genes can emerge in response to physiological and environmental challenges.

- **1. Emergence of Mutations in Iron Metabolism**

- Over the course of evolution, genetic mutations have occurred in genes regulating iron absorption and storage, often in response to environments where iron was either excessive or deficient. Haemochromatosis, for example, is a genetic disease resulting mainly from mutations in the HFE gene located on chromosome 6. This mutation, in particular the C282Y mutation, has been associated with increased iron absorption, leading to iron overload in tissues. The emergence of this mutation could have been advantageous in iron-poor environments, where more efficient absorption would have helped individuals to survive.

- **2. Hereditary transmission and natural selection**

- Genetic mutations such as those affecting the HFE gene are inherited in an autosomal recessive manner, meaning that individuals must inherit two mutated copies of the gene in order to develop clinical disease. In environments where iron levels were variable, these mutations may have conferred a selective advantage on those with an increased capacity to take up iron, but they also led to toxic accumulation in iron-rich environments. This double selective pressure has contributed to the persistence of the mutation in certain populations, even though it can cause serious disorders when expressed at high levels.

- **3. Specific impact on humans**
- Mutations affecting genes linked to iron metabolism are specifically linked to human diseases such as haemochromatosis. In contrast to these mutations, bacteria have developed iron regulation and storage mechanisms that avoid the toxic effects of excess iron, such as siderophores and ferredoxin. These mechanisms enable bacteria to survive in environments with variations in iron availability without the same risks of overload. The absence of similar diseases in bacteria suggests that human mutations in the genes regulating iron are the result of a complex evolution specific to the needs of mammals.

- **Conclusion**

- In conclusion, the evolution of iron regulatory mechanisms in humans and bacteria illustrates adaptation to unique environmental pressures. Genetic mutations such as those associated with haemochromatosis show how changes in regulatory genes can have varied effects depending on environments and physiological needs, with specific implications for human health. The differences in iron regulation mechanisms between bacteria and humans highlight the parallel evolution of biological systems adapted to their particular contexts.

Significance of Low Ferritin and Importance of Medical Consultation

A low ferritin value, such as 20 ng/mL, is often a sign of iron deficiency. Ferritin is an essential protein for storing iron in the body, and its blood levels reflect the iron reserves available. In general, the reference ranges for ferritin are :

- **Men**: 30 to 300 ng/mL
- **Women**: 30 to 200 ng/mL

Ferritin below these ranges indicates iron deficiency, and a level of 20 ng/mL is particularly worrying as it is outside the normal range, suggesting that iron stores are low.

Possible causes of low ferritin

1. **Dietary iron deficiency**: Insufficient consumption of iron-rich foods, such as red meat, legumes and green leafy vegetables, can lead to a drop in ferritin levels.

2. **Impaired absorption**: Conditions such as coeliac disease or gastritis can affect the absorption of iron in the gastrointestinal tract.

3. **Blood loss**: Chronic bleeding, such as that caused by heavy menstruation or gastrointestinal haemorrhage, can lead to a reduction in iron reserves.

Clinical consequences

Low ferritin can lead to iron deficiency anaemia, characterised by symptoms such as fatigue, weakness, paleness and dizziness. In the long term, untreated iron deficiency can affect cognitive and immune function, as well as cardiovascular health.

Assessment and Diagnosis

It is crucial to consult a doctor to assess the underlying causes of low ferritin. Further tests may include :

- **Serum iron**: Measurement of iron circulating in the blood.
- **Total Iron Binding Capacity (TIBC)**: Assesses the capacity of blood to bind iron.

- **Transferrin**: Protein which transports iron in the blood and whose levels increase in the event of iron deficiency.

If genetic mutations such as those in the HFE gene are suspected, a genetic test can be carried out to determine the presence of mutations such as C282Y, associated with haemochromatosis. These tests make it possible to understand whether iron deficiency is due to a genetic condition or to other factors.

Management Plan

A precise diagnosis enables a suitable treatment plan to be put in place, which may include :

- **Iron supplements**: Prescribed to increase iron reserves.
- **Dietary modification**: Including iron-rich foods and strategies to improve iron absorption.
- **Treatment of underlying causes**: For example, treating medical conditions that affect iron absorption or resolving blood loss problems.

In conclusion, low ferritin requires a thorough medical assessment to determine the cause and devise an appropriate treatment. Don't ignore these signs and consult a healthcare professional for advice and follow-up tailored to your situation.

Impact of Iron Deficiency on Blood Pressure

Iron deficiency, often indicated by low ferritin levels, can have a significant impact on blood pressure. This condition, which can lead to iron deficiency anaemia, reduces the blood's ability to carry oxygen efficiently, which can influence blood pressure in a number of ways.

Iron deficiency anaemia and blood pressure
Iron deficiency anaemia is characterised by a reduction in the number of red blood cells and haemoglobin, which can lead to hypoxia (lack of oxygen) in body tissues, including the cardiovascular system. This hypoxia can lead to low blood pressure (hypotension), as the heart struggles to pump blood efficiently due to the reduced total blood volume and oxygen-carrying capacity.

Pathophysiological mechanisms
Iron deficiency can lead to a number of mechanisms that affect blood pressure:

- **Reduced blood volume**: The reduction in the number of red blood cells reduces the total blood volume, which can lower blood pressure.
- **Increased Cardiac Output**: To compensate for the low oxygen transport capacity, the heart can increase cardiac output, which can influence blood pressure values.

- **Vascular changes**: Blood vessels may become less responsive and more susceptible to variations in arterial pressure due to weakness of the vascular walls.

Management and Treatment

It is essential to correct iron deficiency to improve ferritin levels, increase red blood cell count and normalise blood pressure. Treatments may include:

- **Iron supplements**: Prescribed to increase iron reserves.
- **Dietary modification**: Including iron-rich foods and strategies to improve iron absorption.
- **Medical follow-up**: Monitor ferritin levels, blood pressure and other blood parameters to assess the effectiveness of treatment and adjust interventions if necessary.

In conclusion, iron deficiency can influence blood pressure, generally by causing low blood pressure due to iron deficiency anaemia. The mechanisms involve a reduction in blood volume, an increase in cardiac output and vascular changes. Thorough medical assessment and appropriate treatment are essential to correct iron deficiency and manage the impact on blood pressure.

In short, the differences in iron regulation and storage mechanisms between bacteria and humans reveal an evolutionary adaptation to the specific needs of each organism. Bacteria, with their flexible systems such as siderophores and ferredoxin, can survive in a variety of environments with fluctuating iron levels. Humans, on the other hand, depend on a complex iron regulation system involving ferritin and transferrin, and genetic disturbances in these systems can lead to serious pathologies.

An in-depth understanding of these differences may provide valuable insights for the development of targeted treatments and improved management strategies for iron-related diseases. Future research should continue to explore these varied mechanisms to discover new therapeutic approaches and precise biomarkers for iron disorders.

Conclusion

Iron regulation is a complex and fundamental area in the biology of living organisms. Iron storage and transport mechanisms vary considerably between bacteria and eukaryotes, illustrating an evolutionary adaptation to specific needs and different environments. Bacteria, using proteins such as ferredoxin and siderophores such as enterobactin, show a remarkable ability to capture and use iron for vital processes. In contrast, humans rely on sophisticated systems such as ferritin and transferrin to maintain the body's iron balance, prevent deficiency and avoid overload. Genetic diseases linked to iron, such as haemochromatosis, reveal the devastating impact of disturbances in these regulatory mechanisms.

Glossary:

- **Ferritin**: Protein responsible for storing iron in cells, mainly in the liver, spleen and bone marrow.

- **Haemochromatosis**: Genetic disease characterised by an excessive accumulation of iron in body tissues, due to increased iron absorption.

- **Iron deficiency anaemia**: Condition resulting from an iron deficiency, leading to a reduction in the production of red blood cells and causing fatigue and paleness.

- **Oxidative stress**: Cellular damage caused by free radicals and reactive oxygen species, often linked to iron overload.

- **Siderophore**: Molecule secreted by bacteria to capture and transport iron from the environment into the cell.

- **Bacterioferritin**: Iron storage protein in bacteria, structured to store iron safely.

- **Ferredoxin**: Iron-containing protein involved in electron transfer in various metabolic processes.

- **Hepcidin**: Hormone produced by the liver which regulates iron absorption and storage by modulating the activity of ferroportin.

- **Transferrin**: Plasma protein which transports iron in the blood.

- **Gene therapy**: Treatment aimed at correcting genetic mutations that cause disease.

References :

1. Anderson, G. J., & Frazer, D. M. (2017). Iron metabolism and the implications for iron disorders. *Frontiers in Pharmacology, 8,* 19.

2. Andrews, N. C. (1999). Regulation of iron metabolism: Insights from mouse models. *Nature Reviews Molecular Cell Biology, 1*(4), 238-248.

3. Beutler, E. (2004). Iron overload disorders: Diagnosis and treatment. *Hematology, 9*(1), 15-23.

4. Cook, J. D., & Skikne, B. S. (2008). Assessment of iron status. *Best Practice & Research Clinical Haematology, 21*(2), 245-258.

5. Frazer, D. M., & Anderson, G. J. (2014). The regulation of iron homeostasis. *Biochimica et Biophysica Acta (BBA) - Molecular Cell Research, 1843*(12), 2187-2198.

6. Ganz, T., & Nemeth, E. (2012). Iron homeostasis in host defense and inflammation. *Nature Reviews Immunology, 13*(5), 281-292. https://doi.org/10.1038/nri3448

7. Hider, R. C., & Kong, X. L. (2013). Chemistry and biology of siderophores. *Natural Product Reports, 30*(5), 493-518. https://doi.org/10.1039/c3np20022g

8. Johnson, J. L., & Myers, K. (2018). Iron regulation and the role of ferredoxins in bacteria. *Journal of Bacteriology, 200*(10), e00709-17.

9. Krezel, A., & Maret, W. (2007). The biological inorganic chemistry of zinc ions. *Archives of Biochemistry and Biophysics, 463*(2), 254-263. https://doi.org/10.1016/j.abb.2007.05.013

10.Nemeth, E., & Ganz, T. (2014). Regulation of iron metabolism by hepcidin. *Annual Review of Nutrition, 34,* 97-114. https://doi.org/10.1146/annurev-nutr-071813-105439

11.Pietrangelo, A. (2004). Hereditary hemochromatosis: A new look at an old disease. *New England Journal of Medicine, 350*(23), 2383-2397.

12.Torti, S. V., & Torti, F. M. (2002). Iron and cancer: More ore to be mined. *Nature Reviews Cancer, 2*(3), 188-195. https://doi.org/10.1038/nrc744

13.Weiss, G., & Goodnough, L. T. (2005). Anemia of chronic disease. *New England Journal of Medicine, 352*(10), 1011-1023. https://doi.org/10.1056/NEJMra041809

14.Zhang, D., & Liu, G. (2015). The role of transferrin in iron homeostasis and its application in therapeutic strategies. *Journal of Clinical Biochemistry and Nutrition, 57*(2), 107-113. https://doi.org/10.3164/jcbn.15-6

I want morebooks!

Buy your books fast and straightforward online - at one of world's fastest growing online book stores! Environmentally sound due to Print-on-Demand technologies.

Buy your books online at
www.morebooks.shop

Kaufen Sie Ihre Bücher schnell und unkompliziert online – auf einer der am schnellsten wachsenden Buchhandelsplattformen weltweit! Dank Print-On-Demand umwelt- und ressourcenschonend produzi ert.

Bücher schneller online kaufen
www.morebooks.shop

info@omniscriptum.com
www.omniscriptum.com

Printed by Books on Demand GmbH, Norderstedt / Germany